21st CENTURY SAFETY AND PRIVACY™

SMARTPHONE SAFETY AND PRIVACY

DALE-MARIE BRYAN

New York

Published in 2014 by The Rosen Publishing Group, Inc.
29 East 21st Street, New York, NY 10010

First Edition

Library of Congress Cataloging-in-Publication Data

Bryan, Dale-Marie.
Smartphone safety and privacy/by Dale-Marie Bryan.
p. cm. – (21st century safety and privacy)
Includes bibliographical references and index.
ISBN 978-1-4488-9573-1 (library binding)—
ISBN 978-1-4488-9586-1 (pbk.)—
ISBN 978-1-4488-9587-8 (6-pack)
1. Cell phones—Security measures—Juvenile literature. 2. Smartphones—Security measures. 3. Technology—Juvenile literature. I. Bryan, Dale-Marie, 1953-. II. Title.
HQ784.T37 B79 2014
395.59—d23

Manufactured in the United States of America

CPSIA Compliance Information: Batch #S13YA: For further information, contact Rosen Publishing, New York, New York, at 1-800-237-9932.

CONTENTS

INTRODUCTION

The smartphone has forever changed the way people communicate. Not only can friends talk, they can also send messages, share pictures, watch videos, and play games together.

You just used your smartphone to browse the Internet and snag concert tickets for a group that you and your best friend have been dying to see. You swipe to a new screen to text the happy news to her. Minutes later, your message tone pings. She has sent a picture of herself giving you a thumbs-up from a restaurant where she is waiting for her parents.

Did you ever stop to think that a few short years ago that couldn't have happened? Only the invention of the cell phone and its brainier brother, the smartphone, made that instant purchase and friend exchange possible.

But what if, while you were sending the payment information for those tickets, someone captured your mother's credit card or banking information? What if a stalker targeted the exact location where your friend took her picture?

Smartphones are amazing. A phone and minicomputer all in one, they give you options your parents saw only in science-fiction movies. But as with many inventions, smartphones have risks. The fact that smartphones have many computer features makes them prone to similar hazards, such as malicious software and hacking. Smartphones, though, are even more attractive to crooks because they are smaller, people use them more casually, and owners often don't secure them as they do their computers.

As they are with the general public, smartphones are becoming more popular with teens. According to the Pew Research Center, 77 percent of twelve- to seventeen-year-olds in the United States had cell phones in 2011. And 23 percent, or about one in four people in this age group, had a smartphone. As of February 2012, 67 percent of Americans ages eighteen to twenty-four owned a smartphone.

Assuming teen cell phone and smartphone use continues to grow, the chances increase that at some point teens will have their privacy compromised or will even become victims of smartphone-related crimes. The following sections discuss the risks to safety and privacy that smartphones present. They also explore ways to avoid becoming a victim.

CHAPTER ONE

HOW CAN SMARTPHONES BE RISKY?

If time travelers from another era materialized on a busy street, they might think human bodies had morphed. They'd see people with one arm raised, talking to the air with thin, rectangular growths connecting their hands to their ears. Then, to the travelers' surprise, the humans would detach the growths and tap them furiously, oblivious to the world.

It may seem strange to think that people used to have to go home or to a phone booth to contact someone via wired connections. Martin Cooper, general manager at Motorola, symbolically cut those wires on April 3, 1973, when he made the first public cellular telephone call. Eventually, the cell phone allowed people to make contact anywhere they could find a signal.

The next innovation was the Simon Personal Communicator, which was technically the first smartphone. It got its name because its makers at IBM wanted customers to think using it was as easy as playing Simon Says. Arriving on the market in 1994, Simon was a cell phone, fax machine, and pager and included a calendar, clock, calculator, and one game. If you plugged in a memory card, you could have a camera, music, and maps, too. But it cost almost $1,100 for all those features, making it unaffordable for most people.

Computers were fast becoming part of people's everyday lives, allowing them to do more and communicate more easily. As companies made computers smaller and more portable, people began toting both laptops and cell phones. Finally, the two were combined, and the smartphone took center stage as the brightest star in the communication world. But with it came risks.

Martin Cooper of Motorola made the first-ever cellular telephone call to his competitor, Bell Labs, in 1973. The phone weighed a whopping 2.5 pounds (1.1 kilograms).

Smartphone Basics

In order to talk about smartphone safety and privacy issues, it's important to understand some basics about what makes

SMARTPHONES BY THE NUMBERS

Numbers from a Nielsen report and a Pew Research Center study give us an idea of how technology is affecting the way people live.

- Over half of all U.S. cell phones are smartphones.
- In 2012, smartphone users downloaded an average of forty-one apps per phone. Users spent an average of thirty-nine minutes per day using their apps.
- From 2009 to 2011, the median number of daily texts sent by twelve- to seventeen-year-olds rose from fifty to sixty.
- The percentage of teens who talk on a landline daily dropped from 30 percent in 2009 to 14 percent in 2011.
- In 2009, 33 percent of teens met with friends face-to-face outside of school, while in 2011 the number dropped to 25 percent.

smartphones unique pieces of equipment. This includes learning how they and other cell phones transmit signals.

Maybe you've had the experience of using a baby monitor while taking care of a child for a family member or neighbor. Baby monitors are used to hear infants at a distance. They send information by analog signals like a radio. The problem is, using analog signals makes it possible for the neighbors next door to pick up the signals if the monitor is left on.

According to Kevin D. Murray, professional security consultant and author of the book *Is My Cell Phone Bugged?*, early cordless phones used analog signals too, so anyone with a receiver could eavesdrop or decode the signals easily. Neighbors might hear a phone conversation. Thieves staking out potential targets could identify when people weren't at home. Later, the receivers were outlawed, but that didn't stop criminals who could get them illegally.

Then cell phones switched to digital signals, which were more difficult to decipher. Intercepting signals became even harder when they were scrambled or encrypted. But criminals kept up. Even today, no phone conversation is perfectly private, and smartphones present a whole new range of possibilities for the less-than-honest.

What Makes Smartphones So Smart?

Smartphones are cell phones on steroids. They are small computers that allow us to talk, text, send instant messages, browse the Internet, make purchases, do online banking, pay bills, and take and send photos or movies.

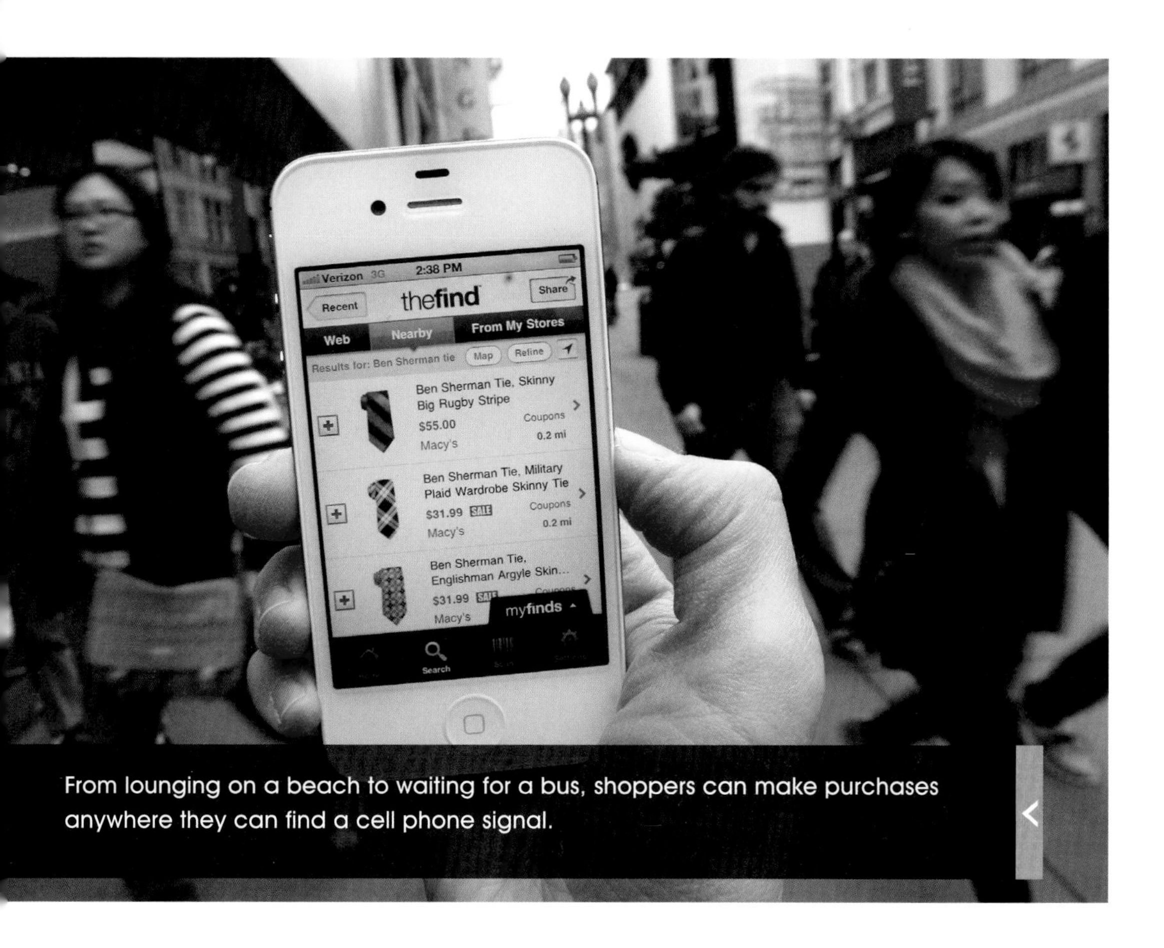

From lounging on a beach to waiting for a bus, shoppers can make purchases anywhere they can find a cell phone signal.

We can download free and low-cost applications (apps) to make our smartphones even smarter. They allow us to find locations, compare prices, play games, read books, and check the weather. They can even become a flashlight, a compass, or a level if we need one. The possibilities are endless, and every day new applications and phone features become available. But it's the smartphone's computer features, and the availability of these enticing additions, that make it susceptible to harm.

Many Kinds of Mischief

As on computers, malware can infect smartphones in order to take information or perform other malicious functions. Some threats that can harm both computers and smartphones include Trojan horses, bots, viruses, and spyware. Each of these threats affects a smartphone in different ways.

Trojan horses are programs that look harmless but contain parts that can hurt the computer functions of your smartphone.

Bots or botnets turn a smartphone into a zombie that reaches out to a boss computer with other smartphone zombies to send unwanted ads or to swamp and shut down Web sites.

Viruses are like Trojan horses in that they sneak in with harmless-looking programs, but then they replicate themselves and infect other computers with the same malicious software.

Spyware is used to collect personal information, often in order to steal people's identities. Some types steal passwords and banking information to take money from unsuspecting victims' bank accounts.

The fact that people love smartphones is a bonanza for cybercriminals. Smartphones are attacked more easily than computers because owners are less likely to protect them with security software. The speed of cyberspace, and the fact that laws have not kept up with our technological age, means cybercriminals are seldom caught. So the problem keeps growing.

Other Serious Issues

Cybercrime is not the only risk using a smartphone poses. Nearly every day we hear about the tragic consequences of using cell phones, especially texting while driving. And distracted walking is a problem, too.

Other dangers of smartphones include cyberbullying, sexting, sextortion, and cheating. All of these can affect the futures of those who use their phones in such irresponsible and harmful ways. In addition, the smartphone's geolocation capabilities allow criminals to use this technology for stalking or spying on people.

People's lives have changed because of the smartphone. Yet it is merely a tool. You are in charge of how you use smartphone technology. It is up to you to be aware of the dangers and to protect yourself by using it responsibly.

CHAPTER TWO

THREATS TO YOUR IDENTITY

Before downloading that game app that looks like so much fun, you might want to give it a second thought. Cybercriminals often use apps to deliver bad software to smartphones. This malware can give criminals access to information on your phone, which they can use to steal your identity.

Identity theft is when someone steals personal information such as a Social Security number, name, birth date, or other personal identifiers without a person's knowledge. Thieves use this information to open credit card accounts to make purchases, rent apartments, or even buy cars and homes. In reality, thieves take a free ride on your name. All the while, negative financial reports stack up against you, which can seriously affect your future.

Smartphones and Identity Theft

According to a study by the Javelin Strategy & Research firm, smartphone users are about 30 percent more likely to

A high school student conducts an identity theft workshop for middle school students in South Carolina. Identity theft is on the rise, and smartphones are making it easier.

become victims of identity theft than nonusers are. The same study showed that nearly twelve million people became victims of identity theft in 2011. This was up 13 percent from 2010. The increase is blamed on the growing use of smartphones and social media.

Cybercriminals are targeting smartphones and social media networks more often because it's becoming harder for them to use people's credit cards illegally. The companies are monitoring customers' card use more closely by calling to check about suspicious charges. To get around this, thieves steal information to get new credit cards in a person's name. That person may be an unsuspecting teen who has never owned a credit card at all!

Just as schools keep track of students' progress, credit agencies collect information about people's money habits. Both the good and the bad are combined in a credit report and are factored together for a credit score, which is like a grade. When people use credit cards wisely and pay their bills on time, they get a higher credit score. The score shows that they make wise choices and pay back borrowed money. Institutions such as banks and colleges use this information to decide if they can safely loan money to you for buying a car or going to school. But if thieves use your identity, they rack up negative reports against your name. Some young people don't even know their identities have been stolen until they apply for a college loan, try to open a bank account, or seek a driver's license.

Teens Are Prime Victims

A 2011 study by the Carnegie Mellon CyLab uncovered disturbing information about the rate of child and teen identity theft. After checking the identities of more than forty thousand children, researchers discovered that more than four thousand (or 10.2 percent) had someone else using their Social Security numbers. The study found that 71 percent of the young victims were between eleven and eighteen years old.

SAFEGUARDING YOUR IDENTITY

It can take years for a victim of identity theft to regain control and erase the damage to a ruined reputation. But you can guard against this crime by staying alert and following some simple steps:

- Be careful what apps you download. Check ratings and comments, and download only apps from proven and safe sources.
- Never respond to unwanted voicemails, e-mails, or text messages. Don't click links that you receive this way, especially if they are from sources you don't recognize or that look suspicious.
- Monitor your credit information. Ask your parents or guardians to help.
- Check credit card statements carefully. Ask your parents to check their statements if you've used their card.
- Ask about adding a security plan to your smartphone. Some computer software security or phone companies offer free or inexpensive plans.

Although following these steps will go a long way toward protecting your identity, it is important to stay informed about new threats so that you can continue to protect yourself in the future.

Children and teens are vulnerable to becoming victims because they have no credit histories. Thieves can use their credit for years because few teens or parents know to check teens' credit reports. Teens may think, "Why worry? I'm only a kid!" and their parents may believe the same.

Unfortunately, identity thieves are hard to catch. And every time authorities become aware of cybercrime schemes, criminals come up with new ones. For example, criminals are figuring out ways to steal information from more people at once. They target servers or databases that store personal information about multiple people. Once their personal information has been stolen, people are 9.5 times more likely to become victims of identity theft, according to Javelin Strategy & Research.

How Do Criminals Steal Information?

Criminals steal information in many ways. They may physically take it by dumpster diving for credit card mailings, bank statements, checks, or tax information. They may steal wallets, purses, or even personal records from places of business. But criminals may also steal information digitally through skimming, keylogging, phishing, smishing, or vishing.

Skimming is when thieves use a special device to copy and store credit card numbers when people use them in restaurants or stores. Sometimes such devices are attached to

Never respond to messages or click links from someone you don't know. It could be a phishing or smishing scam to gain access to your private information.

gas pumps or other places where people swipe their own cards.

In keylogging, special software captures the keystrokes people make when they enter passwords on their phones. Keylogging equipment can also be installed at credit card terminals to capture passwords and code words.

Phishing is when criminals pretend they are companies or other institutions and send spam e-mails or pop-up messages. These messages "fish" for personal information. They try to frighten people by saying their card or account has been suspended. Then they ask their targets to enter personal and credit card information so that they can verify it. Legitimate companies and banks do not do this.

Smishing is phishing through an SMS (short message service) system, or through text messaging. In this method, crooks send text messages as bait, telling people to call a special phone number or click a certain link. When people respond, malware is downloaded to their smartphone system, making personal information available to the crooks.

In vishing, criminals send voicemail messages that ask for personal information. Or they ask people to call back and then pretend to be a bank or government official asking for this information.

All Is Not Lost

While identity thieves do have many ways of deceiving teens, all is not lost. There are several ways you can protect yourself. The easiest and most important is to use your head. Always think of your smartphone as the powerful computer it is. You would never leave your computer lying around for anyone to tamper with. Likewise, you would never leave your computer unprotected by a password. Make sure you care for your smartphone in the same way. Make your passwords difficult for anyone but you to understand, and change your passwords often. And do not share them with anyone except your parents. This includes your friends.

Though the dangers are many, with common sense and care, you can enjoy using your smartphone and guard your identity, too.

MYTHS AND FACTS

Myth: I'm only a kid. Criminals aren't going to bother targeting the data on my smartphone.

Fact: Teens make prime targets for identity theft because they have not yet established credit histories. They are not likely to check their credit information until well after a criminal has used their identities to commit crimes.

Myth: It's best to stay in communication with a cyberbully.

Fact: The only communication you should have with a cyberbully is telling him or her to stop. Then block the person.

Myth: Exchanging passwords with a friend is a great way to show commitment to each other.

Fact: Relationships change all the time. You can never be sure that your password is safe unless you keep it to yourself.

CHAPTER THREE

THREATS TO YOUR SAFETY

A worrisome feature of smartphone technology is the ability to pinpoint the exact locations of users. Anyone carrying a smartphone can easily be found. Say you send your friends a picture of yourself trying on a new outfit at a store. With a free browser add-on, anyone can find out the name and address of that store from the image you sent. In fact, the sophistication of the smartphone's mapping software makes it possible to pinpoint the exact dressing room you are using. So your innocent wish to connect with your friends in a fun and personable way could put you in real danger.

Of course, it's great to see friends' pictures and share locations. Sharing through social networking sites makes people feel that the world is a smaller, more closely knit place. And it is a great way for friends and family to stay in touch. But to those who want to cause trouble, the hidden information attached to photos or videos presents all kinds of dishonest, and even dangerous, possibilities.

What Is Geotagging?

Geotagging is the process of adding global positioning information to photos or videos. Bits of data known as geotags are embedded in photos and videos created with GPS-equipped smartphones. Those who know how to root

A smartphone's GPS technology can provide useful information, such as the locations of nearby stores and restaurants. However, it also allows you to be tracked without your knowledge.

out this information can find the exact degrees latitude and longitude where a picture or video was taken.

Geotagging has its merits. People can check in on their smartphones to get discounts at area stores. They can invite friends to a nearby event or make arrangements to meet at a specific location with picture clues to lead the way.

Geotagging is also a great educational tool. It helps students connect the study of geography to real life. How cool is it to watch real-time video of a rain forest for science or follow the work of archeologists at a dig in Egypt? But as with all innovations, we have to ask if the merits outweigh the downsides.

According to the Privacy Rights Clearinghouse, geotagging makes it possible for people to find out such personal information as your home address, workplace, doctor's office, house of worship, and the route you take to school or work. In some cases, burglars can use geotags to find your house when you are away. This makes your home an easy target.

Advertisers use geotags, too. Knowing the location of your smartphone can help them target their message and their offers more specifically to you. While this may sound like a good thing, gathering this kind of information invades your privacy.

During the 2012 U.S. presidential election, if smartphone users downloaded apps for the two major candidates, Barack Obama and Mitt Romney, they supplied the candidates' offices with information such as the device ID, mobile phone number, GPS location data, contacts, and call logs. According to an article on NetworkWorld.com, the Romney campaign's app requested permission to access users' Facebook pages, as well as camera and audio recordings. The Obama campaign's app offered users information on nearby registered voters and tips on how to convince them to vote their way.

Geotagging not only intrudes on your privacy but also leaves a trail like Hansel and Gretel's bread crumbs. Only instead of birds gathering the bits, others with more commercial, political, or even criminal intentions are doing the pecking. One way to prevent geotagging is to change the settings on your phone. The owner's manual or a phone store representative can show you how to disable the feature when you are not using it.

The First Flames

Cyberbullying is another problem that can start in a small way but mushroom into a serious situation. A girl accidentally bumps into another in the cafeteria, spilling her tray down the second girl's front. Even though she helps her clean up, the second girl is embarrassed and furious. The first girl is surprised by the second's anger. She decides the best thing to do is to let it go and give her time to cool off. She thinks about doing something nice for her later to make amends.

But that night after she finishes her homework, she checks her Facebook page. Someone has posted a comment about her being the world's biggest klutz. She suspects it is the girl from the cafeteria, but the post is tagged with a name she doesn't know. During the week, she finds out about several mean polls that are going around about her. By the end of the month, someone has used her name to subscribe to several pornographic Web sites, and now she is getting all kinds of awful messages and pictures. She also starts getting nasty instant messages at all hours. But she hates to tell anyone, especially her parents. They'd probably take away her computer and her smartphone! Through no fault of her own, she has become the victim of a cyberbully.

In the hands of a cyberbully, a smartphone can become a menacing weapon, inflicting pain with words and pictures.

Cyberbullying and Smartphones

Those first angry texts in the incident were called flaming, and that's exactly what they do. They whip sparks

WAYS TO STAND UP TO A CYBERBULLY

If someone is bullying you online, you are not alone. Don't bring yourself down to the cyberbully's level by bullying back. Here are ways to combat the bullying and seek a resolution:

1. Clearly tell the person to stop. Then don't respond after that.
2. Block the cyberbully's messages. Don't accept messages from anyone you don't know.
3. Tell your parents or another trusted adult.
4. Disable the geotagging function on your smartphone.
5. Make sure your privacy settings on social networking sites are up to date.
6. Report the bully to the Internet service provider (ISP).
7. Call the police immediately if you receive any violent threats. Cyberbullies can and do act on them in person.
8. Keep electronic copies of all the messages you've received, and keep records of the steps you've taken to resolve the situation. Educate yourself. Check out the sources in the For More Information section.

Remember that cyberbullying is not your fault. No one has a right to intimidate you.

ignited by something like the unfortunate cafeteria incident into a full-fledged fire with the help of the Internet. Before long, the activity can take on a life of its own as the flamer delights in the power she feels making the victim miserable. It's like being able to make fun of someone on the playground—but anonymously and with a much bigger audience. If the flaming increases and the flamer is obsessive about it, flaming becomes cyber harassment. When young people do it to each other, it is called cyberbullying.

Bullies may engage in this behavior because they think it's funny or because they want the victim's attention. They may be trying to raise their own status at school, or they may be jealous or seeking revenge. But it is wrong, and teens can take steps against it.

Unfortunately, smartphones make cyberbullying easier by allowing teens to bully victims at any time of day, no matter where they are. A survey by the National Crime Prevention Council showed that 43 percent of teens were victims of cyberbullying in 2011. With the number of smartphone-owning teens increasing each year, smartphones are becoming a powerful tool for cyber cruelty.

But there are ways to tackle the problem. According to the same survey, teens think the best ways to defeat cyberbullying are to block the bullies, refuse to pass along their messages, and tell them to stop.

CHAPTER FOUR

IS YOUR SMARTPHONE DANGEROUS TO YOUR HEALTH?

Smartphones can threaten privacy and security. But they also present physical risks. For one, having a smartphone makes a person a target. Why? Think about it. There you are, glued to a smartphone screen. You are so busy playing a game or watching an episode of your favorite TV show that you don't notice a man watching you from a few feet away on the subway. The car screeches to a stop, and, wham, the thief grabs your phone and flees in seconds.

But consider yourself lucky. According to a report from *CBS News*, the rate of serious crimes involving smartphones is rising. In the United States, 30 to 40 percent of all robberies nationwide involve smartphones or cell phones. And in France, authorities blame the attractiveness of smartphones for a 40-percent increase in violent robberies on the Paris Métro, according to the *New York Times.*

No Laughing Matter

Thieves make money selling stolen smartphones online. Some even joke about it, calling the stealing of iPhones "apple picking" for the company that makes them, Apple. But it is not a joking matter. According to CBS New York, in 2012, a young chef was killed on his walk home from the subway while talking on his iPhone. Later, the thieves tried to sell the phone on Craigslist for $400.

Following common-sense rules can keep you and your smartphone safe. For one, avoid using your smartphone in public places, and keep it out of sight as much as possible. Also, always be aware of your physical surroundings and the people nearby when using a smartphone.

Distracted Driving

One of the most dangerous threats to physical safety involving smartphones is using them while driving. "I need to quit texting because I could die in a car accident," twenty-one-year-old Chance Bothe texted minutes before he drove his truck off a bridge and into a ravine. According to the *Huffington Post*, he broke nearly every bone in his body and had to be revived three times. Following the accident, he needed numerous reconstructive surgeries. Because he suffered brain injuries, he had to learn how to speak again.

Texting and driving don't mix, no matter how competent you are as a driver. Put away your phone, or you risk injuring or killing yourself or others.

The National Safety Council (NSC) estimates that talking on cell phones and texting while driving causes 1.6 million car accidents every year. Yet teens continue to do it. In a 2009 survey by the Pew Research Center, 26 percent of teens ages sixteen to seventeen admitted to texting while driving. Sixty-four percent of teens in this age group said they had been in a car while the driver was texting. And 43 percent said they had talked on a cell phone while driving.

According to research by University of Kansas professor Paul Atchley, while drunk driving increases the risk of having a car crash by 400 percent, talking on a hands-free phone increases it by 500 percent, and texting and driving increases it by 2,300 percent. Atchley told the *Oread* newsletter, "Text messaging while driving is probably the most dangerous thing you can do in a vehicle other than driving with your eyes closed."

Distracted Walking

Distracted walking is a problem, too. A 2012 report by Safe Kids Worldwide showed that during the previous five years, the rate of teens injured while texting and walking had increased by 25 percent. The report related this to the increase in teen cell phone use during the same time period.

According to an article on Mashable.com, in 2011, a security camera filmed a young man on a cell phone falling onto commuter train tracks in Philadelphia. Luckily, no trains came before he was able to climb to safety. In other parts of the country, video cameras caught a cell phone user falling into a fountain at a mall and another walking into the path of a black bear.

A 2012 *Consumer Reports* article said smartphones put pedestrians in a fog. The magazine's survey found that during the past six months, 85 percent of Americans had seen someone walking and using a mobile device to text, talk, use apps, or e-mail. Of that number, 52 percent believed the walkers were putting themselves and others in danger.

If you use your smartphone while walking, you are less likely to notice oncoming vehicles and dangerous obstacles.

Many of those surveyed saw phone users bump into people and walk in front of moving vehicles. With the increase in the number of people owning smart-phones, this risky behavior is bound to increase.

Overuse Injuries

Physical and massage therapists are seeing more patients with aches and pains because of smart-phone use. One ailment, known as "BlackBerry thumb" or "Nintendo thumb," is brought on by long sessions of texting or gaming. Symptoms of the condition include pain, swelling, and stiffness around the base of the thumbs. There are different treatments, including splinting and injections, but the easiest are resting and limiting texting and gaming time.

Neck injuries are on the rise, too. Dr. Dean Fishman, a Florida chiropractor, told CNN he is seeing a growing number of patients with technology-related injuries, including one he calls "text neck." Constantly hunching over while texting, e-mailing, and gaming causes neck and shoulder pain and headaches. It can also lead to disc injury, muscle strain, and pinched nerves, which can lead to arthritis. Usually an old-age condition, teens are developing arthritis from bad posture while using a phone.

Smartphone Addiction

Psychologists are concerned that people are becoming addicted to their smartphones. Dr. Lisa Merlo, director of psychotherapy training at the University of Florida, said she's watched people play with their smartphones rather than look at other people at a party. She told Ellen Gibson of the Associated Press, "The more bells and whistles the phone has, the more likely they are to get too attached."

Gibson also spoke to Michelle Hackman, a high school graduate from Long Island, New York, who won a $75,000 Intel scholarship with her research project about teens' attachment to their cell phones. Most of the teens in her school had smartphones, and she found that their heart rates dropped if they were separated from them. In addition, they didn't know how to stay entertained without them.

According to Dr. Merlo, some people are so attached to their smartphones that they name them and buy them outfits. A *Time* magazine survey showed that 50 percent of

Americans sleep with their phones like teddy bears. Many others keep them right next to their beds. They are afraid they'll miss something if they don't.

But this is not a good practice, Michael Breus, a psychologist and sleep specialist, told Gibson. He said that people who use their smartphones right before bed can't wind down and relax as they need to for a restful night's sleep. Since growing teens need more rest, sleeping with a smartphone could potentially jeopardize their health.

Psychologists are also worried that our attachment to smartphones may be making us less smart. Dr. Susan Krauss Whitbourne wrote in *Psychology Today* that as people rely on their smartphones more, they don't use their brains as much. For example, people no longer have to memorize phone numbers, and they use their smartphone's GPS function rather than a map.

Other Risks

Smartphones can affect your eyes and ears, too. Staring at your smartphone screen for too long can cause a condition called computer vision syndrome, or CVS. Symptoms are dry eyes, headaches, and blurry vision. In most cases, the symptoms are temporary and ease after resting the eyes. However, some individuals experience continued problems with their eyesight.

Smartphone use may also contribute to tinnitus, ringing in the ears that can't be cured. Those who use their phones

> SMARTPHONE HEALTH TIPS

Most people know that not using sports equipment properly can cause injuries. But who would think a piece of communication equipment smaller than a hockey puck could cause a trip to the doctor or emergency room? Following these simple tips could keep you from becoming the next smartphone casualty:

1. Keep your neck straight when using your smartphone. Don't hunch over your device.
2. To reduce exposure to cell phone radiation, use headphones or hold your phone about an inch away from your head while using it.
3. Turn your smartphone off at least one hour before bedtime to allow yourself to unwind.
4. Limit texting time to five-minute sessions.
5. Don't use your phone in the bathroom.
6. Clean your phone frequently using a wipe designed for electronic devices.
7. Avoid using your smartphone in public places. Keep it out of sight.

The best advice for staying healthy when using your smartphone is to use it in moderation and stay focused on what is going on around you.

for long periods of time, or who listen to audio content at high volume, have a greater chance of developing it.

Although there is no solid evidence that radiation from cell phones causes cancer, some experts think there is reason enough to be concerned. In 2011, the World Health Organization (WHO) classified cell phone use as "possibly carcinogenic to humans." It added cell phone radiation to its watch list of possible cancer-causing agents. Some doctors are especially worried about children and young people because their skulls are thinner and their brains are not as well protected.

Smartphones are also germ magnets. According to a study in the *American Journal of Infection Control*, about 40 percent of the cell phones of patients and visitors at a hospital carried harmful bacteria; about 20 percent of the cell phones of health workers did as well. Equally disturbing, a study by marketing agency 11mark revealed that 91 percent of twenty-eight- to thirty-five-year-olds use their smartphones in the bathroom. Hopefully, teens will be wise enough not to follow that trend!

10 GREAT QUESTIONS
TO ASK A PRIVACY AND SAFETY EXPERT

When am I old enough for a smartphone?

How can I protect my identity?

Is there anyone with whom I can share my passwords?

Do I need a special security program for my smartphone?

What do I do if my smartphone is stolen?

What do I do if my smartphone is infected by a virus?

How can I convince my parents to let me get a smartphone?

What if someone is saying mean things about me online?

How can I safely take and send pictures with my smartphone?

Is it possible to repair my online identity?

CHAPTER FIVE

ARE YOU USING YOUR SMARTPHONE IN RISKY WAYS?

Many smartphone threats originate with other people, often strangers. Sometimes, though, teens create their own problems with smartphones. Maybe you have heard about ways to change your phone's operating system so that you can use new, cool apps that you can't get through your phone's manufacturer. What's the harm? Plenty! Doing this can damage your phone and leave you more exposed to privacy and security threats. In addition, there are other risky activities that teens sometimes try with smartphones, including hacking, cheating, and sexting.

Rooting, Jailbreaking, and Hacking

The terms "rooting" for Android phones and "jailbreaking" for Apple phones describe the process of accessing the root file of a smartphone's operating system. This allows a person to change the system's code and add software the manufacturer didn't intend the phone to have. Even though these changes might give you access to a greater number of apps, they can ruin your phone or leave it vulnerable to viruses and other malware.

While changing a smartphone's system is not against the law, hacking is. You may have seen movies in which teens hack into huge computer systems to take them down. Movie producers promote the Robin Hood storyline of the little guy taking down the big guy. But in reality, activities like hacking and copying software are illegal. They can endanger innocent people, including other teens, and land the lawbreakers in jail.

According to an article in *USA TODAY*, the American Psychological Association (APA) surveyed 4,800 teenagers and found that 38 percent had copied software and 18 percent had gone onto another person's Web site or computer without permission. The survey also showed that 13 percent had changed a computer system and 16 percent had taken material that didn't belong to them. Some teens take part in activities like hacking as a way to fit in. Others do it just to see if it can be done. Still others with questionable values think hacking might be a good way to make money. The growth of wireless networks plays a part in this risky behavior because they are more vulnerable to hacking.

Cheating

Cheating with cell phones has become a problem, too, and smartphones make it easier. Gone are the days when students traded slips of papers with answers between classes or let each other copy from tests. Now students who want to cheat can text each other the answers or take pictures of exam pages. At New York's Stuyvesant High School, seventy students were caught using their phones to cheat on citywide exams, according to a 2012 article in the *New York Times*.

Smartphones have become tools for cheating in many schools across the country. But teachers and administrators are becoming savvier about preventing this abuse of technology.

> ARE YOU COMMITTING SMARTPHONE SINS?

Find out by going through this checklist. Then let it be a guide to help you think before you act. Have you ever used your smartphone to...

- Tease or frighten someone?
- Find out something embarrassing or hurtful about a person?
- Send an angry tweet?
- Get someone in trouble in school?
- Track a person down?
- Urge others to be mean to someone?
- Pretend to be someone else?
- Victimize someone to "right a wrong"?
- Feel big or brag?
- Harass someone when you're bored?
- Make threats?
- Send suggestive pictures of yourself?
- Send sexually charged messages?

If you answered "yes" to any of these questions, you are not always using your smartphone responsibly. These behaviors can harm others, and they can harm you. In the future, take more time to think things through before you hit "send."

The same year, another cheating scandal was uncovered at a high school in Texas; sixty students were disciplined.

A poll conducted by the Benenson Strategy Group with Common Sense Media showed some alarming statistics, which were released in 2009. More than a third of teen cell phone users admitted to cheating at least once with them. More than a quarter of those surveyed had used their phones to store information to look at during a test. About a fifth of those surveyed didn't think checking a phone, searching the Internet, or texting friends for answers during a test was wrong.

Educational psychology professor Eric Anderman from Ohio State University studies the cheating issue. In an interview with the *Chicago Tribune*, he estimated that as many as 85 percent of high school students have cheated at least once. Some officials blame it on the increased pressure students feel to make the best grades. Others believe there is no excuse.

Sexting

According to a 2012 report from KCCI News-TV, a girl in Independence, Iowa, used her cell phone to take an obscene photo of herself, which she sent to a boy at her school. He, in turn, sent it to several others who passed it on to their friends. What the girl thought was private, for the boy's eyes only, became very public and permanent. When she gets older and is ready to find a job or a scholarship, that image will likely come back to haunt her.

Using your smartphone for sexting is extremely risky. A photo shared with one person can end up going viral.

A survey conducted by the National Campaign to Prevent Teen and Unwanted Pregnancy and CosmoGirl.com showed that about one in five teens have taken and sent nude or seminude images of themselves. The survey also found that 39 percent of teens have sent sexually suggestive text or e-mail messages. Some sent such images and messages to a boyfriend or girlfriend, while others did it to be flirtatious or simply as a joke. Unfortunately, in too many cases people share these posts, so a teen's "private" photo session and off-color comments make the rounds like the latest strain of flu.

A popular app called Snapchat promises pictures will disappear from cyberspace in ten seconds, leading some teens to think it's a safe way to sext. But it takes less time than that for a person to take a screenshot and post that same picture on Facebook. Then that supposedly safe picture is there for all to see for all time.

Child Pornography and Sextortion

In the Iowa case, the students involved were charged with misdemeanors. But in some cases, taking sexually oriented pictures and videos and sending them on is considered a felony, especially if the people in the pictures are under eighteen. Then it is considered child pornography. Even simply having the pictures on your phone could mean time in juvenile detention. And those who take and send the pictures could be registered as sex offenders, possibly for life.

Taking part in sexting can also put teens in danger of becoming victims of sextortion, which can have frightening lifelong consequences. Extortion is a crime in which a person uses power or threats to get someone else's money. In sextortion, a person threatens to make the victim's sexual images or private sexual information public. Increasingly, teens are becoming victims.

Often predators or pornographers are on the lookout for photos and videos originally sent as part of sexting or bullying. They may also lurk in chat rooms or on social media sites to befriend and trick teens into giving up personal information. Then they use the images and information to prey on and blackmail teens. They may threaten to reveal the images online or to parents and school officials unless a teen sends more explicit photos. Or they may ask for sexual favors as payment. A teen may comply with the criminal's demands out of fear and embarrassment. The threats can escalate unless the victim, a parent, or a concerned friend informs authorities.

CHAPTER SIX

HOW SAFE IS YOUR SMARTPHONE FUTURE?

With all the risks smartphones present, it is important to learn about how to stay safe and maintain your privacy. But trying to tame technology is like trying to control a writhing, growing snake. Every time you think you have it caged, it slithers in a new direction and no longer fits! With technology changing at warp speed and new products continually coming on the market, it can be difficult to keep up. But keeping some key points in mind will help you stay safe, no matter what comes your way.

Never Private, Always Permanent

When using technology, there are two essential ideas to keep in mind. First, nothing is private in today's digital society. Second, everything you put on the Internet is permanent. Richard Guerry, founder of the Institute for Responsible Online and Cell-Phone Communication, believes if people

Put a picture of your grandmother on your background screen as a reminder to watch what you say and do online.

can develop this mind-set, they will be prepared for whatever cybercriminals send their way. So how do we put these ideas into practice?

The smartest defense against the risks of the cyber world is to remember that everyone sees everything. Whatever you post is there for anyone to see, for all time. If something you write or photograph is going to embarrass you or someone else later, don't post it! If you don't think you can remember that, take a picture of your grandmother and use it as your smartphone's wallpaper, or background image, as a reminder. If you wouldn't say it to Grandma and you wouldn't want her to see it, don't post it.

Be Prepared

Always remain alert when using your smartphone. Don't be so involved in a conversation that you don't pay attention to where you are and who is around you. Also, avoid using your smartphone on public transportation or in congested public spaces. It is like hanging a "Steal Me" sign. Keep your phone out of sight in an inside pocket, purse, or backpack. Then keep a hand on that backpack strap. Thieves look for easy targets and will be less likely to challenge a person who looks aware and ready to take action.

No matter how hard you try, though, your smartphone could be stolen. But there are ways to prepare. Many companies that sell smartphones offer security software to protect your information. It is designed to lock your phone down if you report it has been stolen. Some software makes it possible to use the global positioning function to find your phone or to wipe its information remotely if your phone lands in the wrong hands. Talk to your carrier or phone manufacturer about your options.

Be Mindful

Stay aware by thinking about what communications you respond to and create as well. Don't respond to people you don't know. Make sure you understand how to block people you don't want calling or texting. Watch what you say when e-mailing and texting. Remember the screen shows only your words, not your thoughts or facial expressions. The people on the receiving end may not know you are kidding.

BE A SMART SMARTPHONE USER

Your smartphone has amazing features. But it is only a tool. Follow these tips to be a smart smartphone user:

1. Exercise your brain by memorizing and using phone numbers rather than letting your smartphone dial them for you.
2. Turn off the GPS when you are not using it.
3. Give your brain regular workouts by using a map instead of GPS to find a location.
4. If you get a call while driving, let it go to voicemail.
5. If you must talk on a phone while in the car, pull off the road.
6. Never text or check caller ID while driving.
7. If you don't trust yourself not to use your phone while driving, lock it in the trunk until you arrive at your destination.
8. In social situations, put your phone away or turn it off so that you can pay attention and interact with those around you.
9. When walking, keep your phone in your pocket and stay alert. Think about the objects, activity, and people around you.
10. Think about the ways you are using your smartphone. Are you using it, or is it using you?

Taking care to treat others as you would want to be treated translates to cyberspace, too. Respect others by keeping their information, including phone numbers and e-mail addresses, private. Always check before taking and posting anyone's picture with your phone. Think about how you would feel if you were caught in a less-than-attractive pose and the picture appeared all over Facebook the next day. If you don't want that to happen to you, be sure you don't do it to someone else.

Finally, be smart about meeting anyone in person that you've only had contact with online. It's easy for people to pose as someone else and to lie online, so be very cautious. Tell someone where you are going and when. Even better, take someone with you and meet in a very public place.

Beware! Don't Share!

Passwords are the first line of defense in keeping people from accessing the sensitive information on your smartphone. Refer to your owner's manual or the company's Web site to find out how to set up a password to unlock your phone. There are a variety of ways to do it. You can choose which is best for you.

One way is a pattern lock. The phone screen shows an array of dots, and a person can only open it by drawing a specific pattern on the array. The advantage is that it is quick and easy to remember. The disadvantage is that if you make your pattern too easy, a thief can look at your fingerprints on the screen to figure it out. Make sure your pattern crosses over in several places, so that it can't be easily deciphered.

If you choose a standard password, you want to make it easy for you to remember but too complicated for anyone else to figure out. Do not use your name or birth date. Also, try to use words and numbers that only you would know. Make up a phrase and then just use the initials along with different characters and numbers. Try alternating uppercase and lowercase letters. For example, you might say, "On January 22, my favorite kinds of fruit were crisp apples and juicy peaches." So the password would be: o122MfKoFwCa&Jp.

But no matter how creative you are, if you share your password, it will not protect you. It is like opening the door and inviting everyone in. So beware and don't share, ever!

"But what about my best friend?" you may ask. "How about my boyfriend or girlfriend?" Again, don't share, no matter what. We'd like to believe that we can trust everyone. But we may not know people as well as we think we do, and relationships change. With the speed at which information is transmitted today, you don't want to risk making all of your information public.

A practice that has put some teens' privacy at risk is couples exchanging their passwords as a symbol of their devotion. This can be dangerous. What happens to that information if you fall out of love? Would that once-caring person be likely to use your information to get back at you? A person who truly cares about you won't ask you to share something that could put you and your privacy in danger.

Smartphones in the Future

There are reasons to be hopeful about smartphone security in the future. Manufacturers are adding or planning

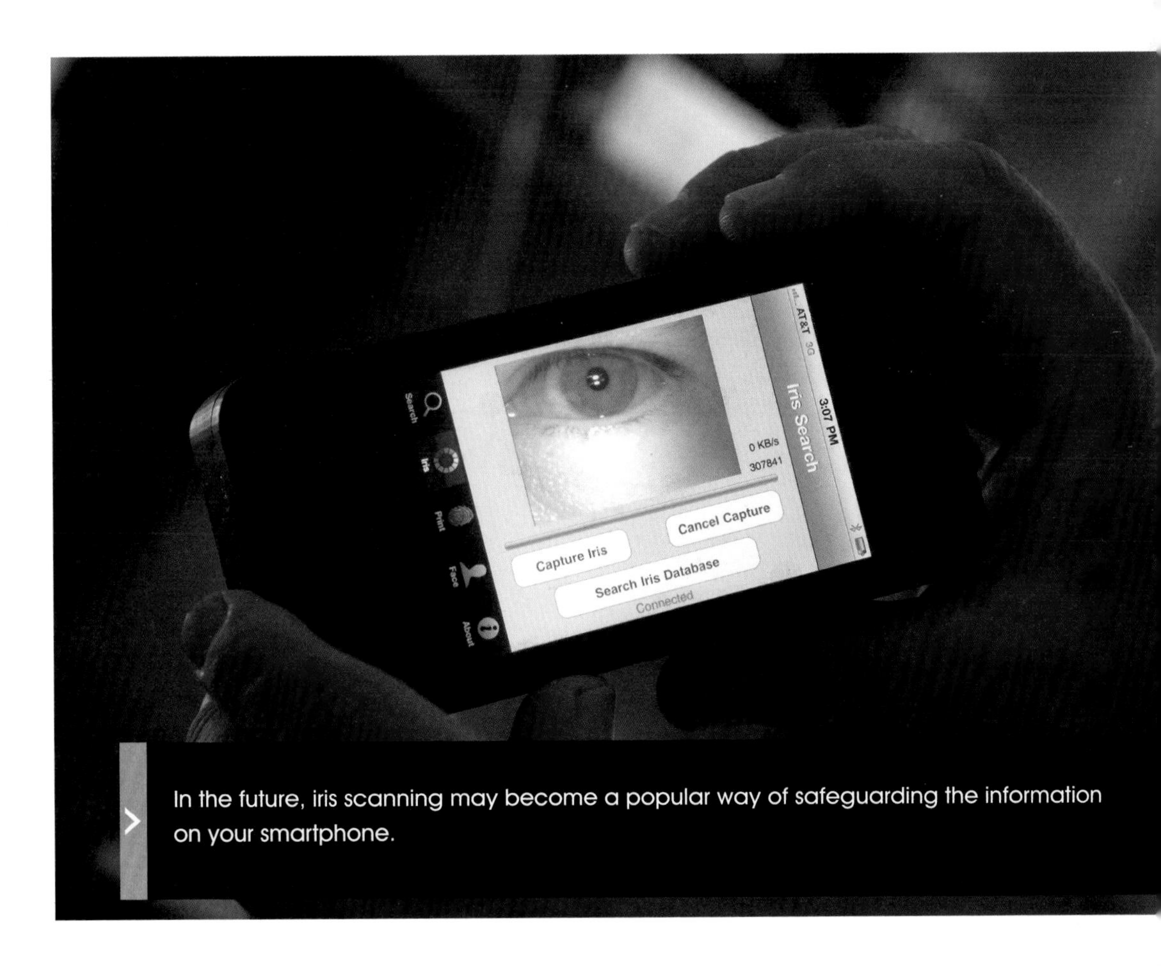

In the future, iris scanning may become a popular way of safeguarding the information on your smartphone.

new features to keep you safer. In the future, unlocking your phone may require scanning your eye, face, or ear or using voice recognition and keywords to do the job. You may have to enter a video password that requires you to look into your phone's camera and speak. Or you may use a combination of these for maximum security.

Companies will provide stronger antivirus software for smartphones and other devices. App stores will become safer. Yes, there are dangers to our privacy and security because of smartphones. But with common sense and continual awareness, we can use and enjoy them safely.

app A software application that is added to a smartphone for greater functionality.

computer vision syndrome (CVS) A collection of eye-related problems associated with the use of digital devices. Symptoms include eyestrain, dry eyes, eye redness, blurry vision, and headaches.

cyberbullying The targeting and harassment of a young person by another using the Internet, e-mail, texting, instant messages, or other electronic communications.

cyber harassment The targeting and harassment of an adult by another adult using the Internet, e-mail, texting, instant messages, or other electronic communications.

flaming The act of writing angry or insulting messages to another online.

geotagging The process of adding Global Positioning System (GPS) data to digital photos or videos.

Global Positioning System (GPS) A navigation system that uses satellite signals and a receiver to determine the position of a person, vehicle, etc.

identity theft The capture and use of someone's personally identifying information, such as a Social Security number or birth date, to commit fraud or other crimes.

jailbreaking Modifying an electronic device, especially an iPhone or other Apple product, in order to remove the restrictions that prevent it from running unofficial software.

keylogging A process in which special software or equipment captures all the keystrokes entered on a computer, smartphone, or other device, usually in order to steal account numbers, passwords, etc.

malware Harmful software that causes undesirable changes in the functioning of a computer or smartphone.

phishing The process of obtaining personal data by posing as a legitimate bank or institution and sending official-looking e-mail messages to "fish" for information from an account holder.

sexting The act of sending photos or messages with sexual content, primarily between cell phones.

skimming The process of using a special device to illegally copy and store people's credit card numbers.

smishing A technique in which thieves send text messages designed to get smartphone users to reveal personal information.

social networking The use of a social Web site or service to communicate and share information with others in one's network.

text neck A condition marked by neck and shoulder pain and headaches that is caused by constantly hunching over a mobile device.

vishing A technique in which criminals send voicemail messages in an attempt to get smartphone users to reveal personal information.

Federal Communications Commission (FCC)
445 12th Street SW
Washington, DC 20554
(888) CALL-FCC [225-5322]
Web site: http://www.fcc.gov
The Federal Communications Commission regulates interstate and international communications by radio, television, wire, satellite, and cable. It works to combat smartphone and data theft and offers information on protecting smartphones and other smart devices.

Identity Theft Resource Center (ITRC)
P.O. Box 26833
San Diego, CA 92196-6833
(888) 400-5530
(858) 693 7935
Web site: http://www.idtheftcenter.org
The ITRC is a nonprofit organization dedicated to the understanding of identity theft and related issues. It provides victim and consumer support, as well as public education. The Web site has helpful information about how to prevent identity theft and where to get help in case it occurs.

National Crime Prevention Council (NCPC)
2001 Jefferson Davis Highway, Suite 901

Arlington, VA 22202-4801
(202) 466-6272
Web site: http://www.ncpc.org
The National Crime Prevention Council's mission is to help people keep themselves, their families, and their communities safe from crime. It offers resources on a variety of topics, including cell phone safety, Internet safety, cyberbullying, and identity theft.

National Organizations for Youth Safety (NOYS)
7371 Atlas Walk Way, #109
Gainesville, VA 20155
(828) FOR-NOYS [367-6697]
Web site: http://www.noys.org
This coalition of youth-serving organizations works to save lives, prevent injuries, and promote safe and healthy lifestyles among all youth. It offers programs and information on topics such as bullying and distracted driving.

PACER's National Bullying Prevention Center
8161 Normandale Boulevard
Bloomington, MN 55437
(800) 537-2237
Web site: http://www.pacerteensagainstbullying.org
PACER is a group that provides children of all ages and adults with information and resources to combat bullying.

Its teen Web site is a place for middle and high school students to find ways to address bullying, to be heard, and to take action.

Partners for Safe Teen Driving
Prince William Network
Prince William County Public Schools
P.O. Box 389
Manassas, VA 20108
(800) 609-2680
Web site: http://www.safeteendriving.org
Partners for Safe Teen Driving is a community health initiative aimed at reducing the number of teenage automobile crashes, injuries, and fatalities. The organization educates teenage drivers and their parents about safe and responsible driving.

Web Sites

Due to the changing nature of Internet links, Rosen Publishing has developed an online list of Web sites related to the subject of this book. This site is updated regularly. Please use this link to access the list:

http://www.rosenlinks.com/21C/Phone

FOR FURTHER READING

Allen, Kathy. *Cell Phone Safety* (Tech Safety Smarts). Mankato, MN: Capstone Press, 2013.

Kiesbye, Stefan. *Distracted Driving* (At Issue). Detroit, MI: Greenhaven Press, 2012.

Klein, Rebecca T. *Frequently Asked Questions About Texting, Sexting, and Flaming* (FAQ: Teen Life). New York, NY: Rosen Publishing, 2013.

Kling, Andrew A. *Cell Phones* (Technology 360). Farmington Hills, MI: Lucent Books, 2010.

Netzley, Patricia D. *How Does Cell Phone Use Impact Teenagers?* (In Controversy). San Diego, CA: ReferencePoint Press, 2013.

Obee, Jenna. *Social Networking: The Ultimate Teen Guide* (It Happened to Me). Lanham, MD: Scarecrow Press, 2012.

Ostow, Micol. *What Would My Cell Phone Do?* New York, NY: Speak, 2011.

Szumski, Bonnie, and Jill Carson. *Are Cell Phones Dangerous?* (In Controversy). San Diego, CA: ReferencePoint Press, 2011.

Vacca, John R. *Identity Theft* (Cybersafety). New York, NY: Chelsea House, 2011.

Wilkinson, Colin. *Mobile Platforms: Getting Information on the Go* (Digital and Information Literacy). New York, NY: Rosen Publishing, 2011.

Alterman, Elizabeth. "As Kids Go Online, Identity Theft Claims More Victims." CNBC.com, October 10, 2011. Retrieved October 2012 (http://www.cnbc.com).

Baker, Al. "At Top School, Cheating Voids 70 Pupils' Tests." *New York Times*, July 9, 2012. Retrieved October 2012 (http://www.nytimes.com).

Common Sense Media. "Hi-Tech Cheating: What Every Parent Needs to Know." 2009. Retrieved January 15, 2013 (http://www.commonsensemedia.org).

Consumer Reports. "Phones Put Pedestrians in a Fog." August 2012. Retrieved January 13, 2013 (http://www.consumerreports.org).

Dellorto, Danielle. "WHO: Cell Phone Use Can Increase Possible Cancer Risk." CNN.com, May 31, 2011. Retrieved January 13, 2013 (http://www.cnn.com).

Dunn, John E. "Obama and Romney Election Apps Suck Up Personal Data, Research Finds." NetworkWorld.com, August 21, 2012. Retrieved October 2012 (http://www.networkworld.com).

Elias, Marilyn. "Most Teen Hackers More Curious Than Criminal." *USA TODAY*, August 19, 2007. Retrieved January 14, 2013 (http://usatoday30.usatoday.com).

Erlanger, Steven. "Smartphones Lure Sticky Fingers in Paris." *New York Times*, January 8, 2011. Retrieved July 2012 (http://www.nytimes.com).

Fitzgerald, Britney. "Chance Bothe Texts 'I Need to Quit Texting' Before Near-Deadly Truck Crash." HuffingtonPost .com, August 3, 2012. Retrieved August 2012 (http://www.huffingtonpost.com).

Freeman, Kate. "Man Falls on Train Tracks While Talking on Cell Phone." Mashable.com, July 31, 2012. Retrieved January 2013 (http://mashable.com).

Gibson, Ellen. "Are Smartphones Becoming an Obsession?" *Newsday*, August 1, 2011. Retrieved January 13, 2013 (http://www.newsday.com).

Gilbert, Jason. "Smartphone Addiction." HuffingtonPost.com, August 16, 2012. Retrieved January 13, 2013 (http://www.huffingtonpost.com).

Glynn, Casey. "iPhone-Related Crime on the Rise as Smartphones Gain Popularity." CBSNews.com, September 12, 2012. Retrieved October 2012 (http://www.cbsnews.com).

Goldman, David. "Half of U.S. Cell Phones Are Now Smartphones." CNNMoney.com, May 16, 2012. Retrieved September 2012 (http://money.cnn.com).

Identity Theft Resource Center. "ITRC Fact Sheet 144—Smartphone Safety." IdTheftCenter.org, August 2011. Retrieved July 2012 (http://www.idtheftcenter.org).

KCCI-TV Des Moines. "6 Teens Charged in Sexting Case." KCCI.com, November 21, 2012. Retrieved January 15, 2013 (http://www.kcci.com).

Keilman, John. "Teachers Put to Test by Digital Cheats." *Chicago Tribune*, August 7, 2012. Retrieved October 2012 (http://articles.chicagotribune.com).

Kelly, Tara. "Toilet Texting, Bathroom Browsing on the Rise, Study Says." HuffingtonPost.com, February 1, 2012. Retrieved January 13, 2013 (http://www.huffingtonpost.com).

Lenhart, Amanda. "Teens, Smartphones & Texting." Pew Internet & American Life Project, March 19, 2012. Retrieved July 2012 (http://www.pewinternet.org).

Lipka, Mitch. "Rise in Identity Fraud Tied to Smartphone Use." Reuters.com, February 22, 2012. Retrieved September 2012 (http://www.reuters.com).

Mueller, Ann Tracy. "Does Your Mobile Device Have Cooties?" Ragan's Health Care Communication News, December 29, 2011. Retrieved January 13, 2013 (http://www.healthcarecommunication.com).

Murray, Kevin D. *Is My Cell Phone Bugged? Everything You Need to Know to Keep Your Mobile Conversations Private.* Austin, TX: Emerald Book Company, 2011.

National Campaign to Prevent Teen and Unplanned Pregnancy. "Sex and Tech: Results from a Survey of Teens and Young Adults." 2008. Retrieved January 15, 2013 (http://www.thenationalcampaign.org/sextech/PDF/SexTech_Summary.pdf).

National Safety Council. "NSC Estimates 1.6 Million Crashes Caused by Cell Phone Use and Texting."

January 12, 2010. Retrieved January 13, 2013 (http://www.nsc.org).
Safe Kids Worldwide. "Walking Safely: A Report to the Nation." August 2012. Retrieved January 2013 (http://www.safekids.org/assets/docs/safety-basics/safety-tips-by-risk-area/Walking-Safely-Research-Report.pdf).
Sager, Ira. "Before iPhone and Android Came Simon, the First Smartphone." *Businessweek*, June 29, 2012. Retrieved October 2012 (http://www.businessweek.com).
Smith, Josh. "Is Your iPhone Trying to Kill You?" GottaBeMobile.com, December 29, 2011. Retrieved January 13, 2013 (http://www.gottabemobile.com).
University of Kansas. "Professor Profile: The Dangers of Texting and Driving." *The Oread*, August 23, 2010. Retrieved January 13, 2013 (http://oread.ku.edu).
Whitbourne, Susan Krauss. "Your Cellphone May Be Making You...Not Smart." *Psychology Today*, October 18, 2011. Retrieved August 2012 (http://www.psychologytoday.com).
Wilson, Charles. "Feds: Online 'Sextortion' of Teens on the Rise." NBCNews.com, August 15, 2010. Retrieved October 2012 (http://www.msnbc.msn.com).
Wilson, Jacque. "Your Smartphone Is a Pain in the Neck." CNN.com, September 20, 2012. Retrieved January 13, 2013 (http://www.cnn.com).

About the Author

Dale-Marie Bryan is a curriculum writer and the author of eleven books for young people. She loves her smartphone but always keeps the rule "nothing is private and everything is permanent" in mind.

Photo Credits

Cover (figure) Ron Levine/Photodisc/Getty Images; cover (background) Peter Macdiarmid/Getty Images; p. 4 Nick David/Taxi/Getty Images; pp. 7, 10 © AP Images; p. 14 © Augusta Chronicle/ZUMA Press; p. 18 iStockphoto.com/Michael Bodmann; p. 22 Christian Science Monitor/Getty Images; p. 25 Nathan Blaney/Photodisc/Getty Images; pp. 30, 43 iStockphoto/Thinkstock; p. 32 Fabrice LeRouge/ONOKY/Getty Images; p. 40 iStockphoto/Nasowas; p. 46 (portrait) Vanessa Gavalya/The Image Bank/Getty Images; p. 51 Boston Globe/Getty Images.

Designer: Michael Moy; Editor: Andrea Sclarow Paskoff; Photo Researcher: Amy Feinberg